ENVIRONMENTAL MANAGEMENT SYSTEMS AND ISO 14001

SUMMARY REPORT

Federal Facilities Council Report

No. 138

NATIONAL ACADEMY PRESS
Washington, D.C. 1999

NOTICE

The Federal Facilities Council (FFC) (formerly the Federal Construction Council) is a continuing activity of the Board on Infrastructure and the Constructed Environment (BICE) of the National Research Council (NRC). The purpose of the FFC is to promote continuing cooperation among the sponsoring federal agencies and between the agencies and other elements of the building community in order to advance building science and technology—particularly with regard to the design, construction, acquisition, evaluation, and operation of federal facilities. Currently, the following agencies sponsor the FFC:

Department of the Air Force, Office of the Civil Engineer

Department of the Air Force, Air National Guard

Department of Energy, Office of Field Management

Department of the Navy, Naval Facilities Engineering Command

Department of State, Office of Foreign Buildings Operations

Department of Veterans Affairs, Office of Facilities Management

Food and Drug Administration

General Services Administration, Public Buildings Service

Indian Health Service

National Aeronautics and Space Administration, Facilities Engineering Division

National Institutes of Health

National Institute of Standards and Technology, Building and Fire Research Laboratory

National Endowment for the Arts, Design Arts Program

National Science Foundation

Smithsonian Institution, Office of Facilities Services

U.S. Public Health Service, Office of Management

U.S. Postal Service, Facilities Department

As part of its activities, the FFC periodically publishes reports that have been prepared by committees of government employees. Since these committees are not appointed by the NRC, they do not make recommendations, and their reports are considered FFC publications rather than NRC publications.

For further information on the FFC program or FFC reports, please write to: Director, Federal Facilities Council, Board on Infrastructure and the Constructed Environment, 2101 Constitution Avenue, N.W., Washington, D.C. 20418 or send an e-mail to lstanley@nas.edu.

Contents

1	INTRODUCTION	1
	Background and Purpose	1
	Organization of this Report	2
	Workshop Structure	2
	Recurrent Themes and Issues	3
2	ENVIRONMENTAL MANAGEMENT SYSTEMS/ISO 14001	6
3	USE OF ENVIRONMENTAL MANAGEMENT SYSTEMS AND ISO 14001 IN THE PUBLIC SECTOR	10
	U.S. Environmental Protection Agency	10
	U.S. Environmental Protection Agency	12
	U.S. Department of Energy	15
	U.S. Department of Defense	18
	California Environmental Protection Agency	21
	U.S. Postal Service	25
	U.S. Air Force	28
4	LOCKHEED MARTIN CORPORATION'S PERSPECTIVE ON EMSS AND ISO 14001	32
APPENDIX A	Biographical Sketches of the Speakers	39
APPENDIX B	Assistance and Resource Documents	44
APPENDIX C	DOD Pilot Study EMS Installations	45

1

Introduction

BACKGROUND AND PURPOSE

The International Organization for Standardization (ISO) defines an environmental management system (EMS) as that "part of the overall management system that includes organizational structure, planning activities, responsibilities, practices, procedures, processes and resources for developing, implementing, achieving, reviewing and maintaining the environmental policy" (ISO, 1996). An EMS enables private companies, federal and state agencies, and other organizations to establish, and assess the effectiveness of, procedures to set environmental policy and objectives, achieve compliance, and demonstrate such compliance to others.

International standards covering EMSs provide organizations with the elements of an EMS that can be integrated with other management functions to help them attain environmental and economic goals. ISO 14000 is a family of standards intended to support environmental protection and prevent pollution in balance with socioeconomic needs. The international standard does not establish absolute requirements for environmental performance beyond commitment to compliance with applicable legislation and regulations and to continual improvement. Thus, two organizations carrying out similar activities but having different environmental performance may both comply with ISO 14000 requirements.

ISO 14000 encompasses 16 standards that address organizational issues and products. ISO 14001 is the EMS specification document outlining the requirements that an organization must meet for its EMS to be registered or certified to the standard. It is a tool to measure the effectiveness of environmental management programs. ISO 14001 is being used by private corporations to enhance their competitiveness in both foreign and domestic markets. Federal and other public agencies are evaluating the potential of ISO 14001 or alternative EMSs to improve performance, and some agencies have already launched pilot projects.

In 1996, the Federal Facilities Council (FFC), which operates under the aegis of the National Research Council, established a standing committee on Environmental Engineering with the express purpose of providing a forum where federal environmental engineers and program managers could meet on a

INTRODUCTION

regular basis to exchange information about facilities-related environmental programs, policies, and issues. The committee members, like environmental program managers in other types of organizations, are increasingly concerned about achieving and demonstrating sound environmental performance by meeting the requirements of environmental regulations and limiting the impacts of their products or services on the environment. To foster communication and address concerns about EMSs, the FFC Standing Committee on Environmental Engineering hosted a one-day workshop on *Environmental Management Systems and ISO 14001*. The workshop was held April 9, 1998, at the National Academy of Sciences in Washington, D.C.

ORGANIZATION OF THIS REPORT

The next two sections of this report describe the structure of the workshop and summarize recurrent themes and issues that emerged during the course of the day. Narrative summaries of each presentation follow. Appendix A contains biographical sketches of the workshop's speakers. Appendix B contains a list of assistance and resource documents. Appendix C includes a list of Department of Defense (DOD) EMS pilot study sites.

WORKSHOP STRUCTURE

The *Workshop on Environmental Management Systems and ISO 14001* featured nine highly qualified speakers who shared their experiences with EMSs and ISO 14001 with an audience of approximately 50 federal, other public-sector, and private-sector environmental managers. George Glavis, chair of the FFC Standing Committee on Environmental Engineering, welcomed the speakers and participants to the workshop, described its purpose, and provided background information on the economic, social, and regulatory climate in which ISO 14001 and EMSs are to be implemented. He then introduced the keynote speaker, Joseph Cascio, lead U.S. delegate to the International Organization for Standardization and vice president of the Global Environment and Technology Foundation. Mr. Cascio gave an overview of EMSs and the principles of ISO 14001.

A panel discussion on the "Use of Environmental Management Systems and ISO 14001 in the Public Sector" was moderated by Mary McKiel, director of the U.S. Environmental Protection Agency (EPA) Standards Network. Dr. McKiel discussed the accreditation program associated with ISO 14001, pointing out that the use of the standard is completely voluntary, as is registration or certification. The panel featured Sarah Walsh, project manager with the Federal Facilities Enforcement Office, EPA; Larry Stirling, Senior Environmental Protection Specialist, Department of Energy; Rick Drawbaugh, representing DOD, and Robert Stephens, Deputy Director for Science,

INTRODUCTION

Pollution Prevention, and Technology Program, California Environmental Protection Agency.

A second panel discussion titled "Implementing Environmental Management Systems/ISO 14001," was moderated by Terry Christensen of the Public Health Service. Speakers included John Bridges, Environmental Compliance Coordinator, Capitol Metro Region, U.S. Postal Service; Stephen Evanoff, Manager, External Affairs, Corporate Environment Safety and Health, Lockheed Martin Corporation; Norman Varney, Jr., Associate General Counsel, Lockheed Martin Electronics; and Rick Drawbaugh, Deputy for Environment, Safety, and Occupational Health Technology, Office of the Deputy Assistant Secretary for the Air Force.

A plenary session on "Issues Related to the Use of ISO 14001 and Other Environmental Management Systems in Public Agencies," was moderated by Catherine Fairlie of the Air National Guard.

RECURRENT THEMES AND ISSUES

Throughout the workshop, members of the audience were given opportunities to question the speakers about issues related to federal regulatory compliance, EMSs, and ISO 14001. Neither the speakers nor members of the audience were asked to come to any consensus on the issues or recommendations for resolving them. However, over the course of the workshop a number of recurrent themes and issues emerged.

Changing the Environmental Protection Paradigm

Traditionally, environmental protection staff within an organization have focused primarily on complying with environmental regulations, a set of standards that are point-in-time concepts relating to environmental performance and protection. Speakers from both the private and public sectors discussed the importance of shifting the environmental protection paradigm from one focused solely on complying with federal regulations to one for which compliance is achieved as part of a more proactive performance-based system. Some organizations discovered that when it comes to environmental protection, compliance is not enough; an organization can be in full compliance with the law and still produce products and services that have negative impacts on the environment.

EMSs were discussed as a means to provide standardized frameworks from which individualized performance criteria can be established and measured. Thus, an organization can know whether it has actually reduced levels of resource consumption or environmental emissions rather than whether it has simply met a regulation. One speaker noted that the change in the environmental paradigm raises the question of how to train people to meet these environmental challenges.

Characteristics of EMSs and ISO 14001

Speakers representing federal and state agencies as well as private companies found ISO 14001 to be a flexible, baseline approach that can be adapted to organizations of all sizes and types, and to a variety of cultures, processes, and businesses.

The fundamental principles of ISO 14001 promote a program of continual improvement to achieve objectives and targets set by the organization itself. The objectives and targets derive from an assessment of significant environmental aspects and commitments made in the organization's policy. The Code of Environmental Principles (CENT), another type of environmental management system, also was described. The CENT specifically addresses compliance, assurance, and pollution prevention, a distinguishing difference from ISO 14001.

Incorporating EMS into Corporate Business Practices

As an EMS is integrated into an organization's business decision-making processes, it can improve program management and enhance environmental performance. Some speakers' organizations found that the EMS improved overall productivity. As an EMS is incorporated into central management systems and organizational strategies, there is less need for external oversight as core business operations take ownership of environmental responsibility.

The full cost of implementing an EMS includes the salary and time costs of in-house staff devoted to the project, as well as costs of any certification program. Total costs will vary, depending on the quality of the organization's existing environmental management/protection systems. Several speakers concluded that an EMS, once implemented, should result in cost savings over time.

One speaker from a large corporation stated that, with the implementation of a corporatewide management system, his organization's long-term vision includes facility self-governance, incorporating self-auditing of their compliance with environmental regulations.

Several speakers addressed the benefit of dealing with environmental, safety, and health issues together as part of business planning and operational risk assessment. They agreed that integrating these issues into the same management concepts makes good business sense.

Changing Organizational Culture

Perhaps the most challenging aspect facing the organizations represented was instituting new management systems that required major change in their internal culture, away from compliance-based reporting and toward more active environmental stewardship. Outreach, education, and training are critical elements in making these programs successful because changing the culture requires analysis of how things were done in the past and what was wrong with that approach, as well as why the new system is better. It also means establishing performance measures to facilitate change.

One public agency has a number of EMS-related education programs under way. It is providing training and sharing information through the publication of fact sheets. An EMS topical committee is being established to disseminate information and provide a forum for discussion. Technical assistance has been provided to a few agency sites and a World Wide Web site has been developed.

To shift from environmental compliance to proactive environmental protection and prevention of pollution, organizations must involve all of their employees, build an infrastructure to support them in taking responsibility for the environmental aspects of their jobs, and educate them about environmental issues. One key to the successful implementation of ISO 14001 or another EMS is that these systems require people to get involved at all levels of the organization, from the top management down. Individual responsibility and accountability at the employee level is emphasized.

REFERENCE

ISO (International Organization for Standardization). 1996. ISO 14001: Environmental Management Systems—Specification with Guidance for Use, No. ISO 1996 (E). Geneva, Switzerland: ISO.

2

Environmental Management Systems/ISO 14001

Joseph Cascio
Vice President
Global Environment and Technology Foundation

What is different about ISO 14000 and why is there some hesitation about accepting it in industry? ISO 14000 is different from other environmental standards and initiatives because of its focus. Other programs have concentrated on what is important to the regulatory agencies and thus have focused on environmental compliance. ISO 14000 addresses and expects compliance, but its focus is on building an infrastructure within an organization that will address compliance, improve performance, and achieve environmental objectives.

Traditionally, environmental protection departments within organizations have not focused on protecting the environment; rather they concentrate on complying with the law. To make a significant difference, organizations must change focus; get employees involved; build the infrastructure to support them; and make people aware, competent, understanding, and knowledgeable about environmental issues. Individuals must take responsibility for environmental protection within functional units and the line management also must be responsible if results are to be achieved.

To illustrate, when IBM began a solid-waste recycling program, people were sensitive to environmental issues, but the project looked big and intimidating. At the time, IBM employed 425,000 people. The question was how to motivate them all.

The IBM team created a new logo for solid-waste recycling. Mugs and T-shirts were distributed to all employees, building on a motivational approach. The corporation was not telling them to do something for itself, such as work harder, make a better product, or make more profits, but to help protect the environment. From the employees' point of view, they were involved not for the company, but for themselves, society, Earth, and nature. It was easy to get them motivated.

ISO 14000 emphasizes building the capacity of the organization to address its environmental issues, to comply. More than that, the standard aims to invigorate the employee population, to get them excited about the program, and to achieve savings and efficiencies. An organization is going to implement

ISO 14000 because it sees other benefits accruing beyond compliance with environmental regulations.

Fundamental Principle Of ISO 14000

The fundamental principle of ISO 14000 is continual improvement. It does not matter where an organization starts; as long as it starts, its overall goal will be continual improvement. ISO 14000 helps an organization to build structural capacity toward a logical framework to address environmental compliance, improve performance, and achieve other established environmental objectives, such as preventing the creation of pollution.

ISO 14000 is complementary to national environmental regulations. It borrows the quality management standard from ISO 9000, which served as a model for the ISO 14000 internal structure. The old philosophy in quality management, for example, was to focus on defects, to stress the reduction of errors in employees' work, and to make a better product. The emphasis was constantly on the quality of the product. It became evident that this approach brings only marginal improvements. A culture of quality must be created and embedded in an organization's processes and people to produce a quality product.

The philosophy of enriching and improving the structure of the system came from the quality movement, and now has been applied to the environmental area with ISO 14000.

ISO 14000 encompasses 16 standards, in addition to that on environmental management. Six standards address organizational issues and ten address products. They include environmental auditing, performance evaluation, environmental labeling, development, and site assessment. The ISO 14031 document, for example, addresses performance evaluation. It does not contain specific performance indicators, but it establishes a framework for determining the attributes needed in choosing performance indicators and establishing a subsystem for performance evaluation.

ISO 14001

ISO 14001 is the specification document. An organization must implement the elements of this specification to be registered to the standard.

In reading the section in the ISO 14001 standard that pertains to compliance, one might think it is deficient because it says very little about compliance, but, it is not deficient. The standard requires an organization to have a policy that explicitly states that a commitment has been made to compliance, to continual improvement, and to prevention of pollution. To be registered to the ISO 14001 standard, an organization must have a procedure and a process for determining the legal requirements that apply and must be met.

ISO 14001 has a requirement for setting objectives and targets in an organization's management system for all significant environmental aspects and for the commitments made in the policy. Once an organization does so, the standard states that it must create an environmental program to achieve those objectives and targets.

To attain these objectives, ISO 14001 expects organizations to train their employees, to allocate resources, and put operation controls in place to attain and remain in compliance. Line management must take ownership of the requirement to achieve the objective by getting employees involved. The environmental department staff no longer stands alone, for the line management has responsibility and ownership of objectives and targets, and employees are directly involved.

Organizations need a methodology and a structure in place to succeed in these areas. One requirement is a periodic check on compliance status. In effect, this means that compliance auditing must be conducted. Also, compliance must become part of a management review process. The purpose is to propel continual improvements and allow the organization to gauge whether it needs to apply more resources or make changes in the system to achieve compliance.

ISO 14001 is not a regulation, but a voluntary standard. It is a system that provides the structure and tools to permit an organization to do the best job possible in meeting environmental management goals.

Do organizations want to continually improve performance? Or do they want to continually improve the management system, structure, ability, and capacity to actually address the issues? The requirement is that organizations continually look at the system and improve it so that capacity to achieve performance goals and compliance can be improved.

When capacity is improved, the organization can decide how to use it. Hopefully, that capacity will be used to improve environmental performance. Because it is part of the management review process, top management is involved as never before and that involvement creates the impetus for continual improvement.

What does ISO 14001 imply for organizations? That they now accept responsibility and take ownership of environmental issues. Compliance is part of the system but the organization is going to identify the objectives and targets it will meet, the resources it will apply, the programs it will implement, and the training it will provide.

How are the terms of success defined? By the organization itself. The organization sets objectives and ISO 14001 is used to achieve those objectives. Organizations that do so will actualize savings and efficiencies and avoid liabilities and embarrassment.

ISO 14001 fosters understanding, esprit de corps, and commitment. It facilitates changing the internal culture of the organization, so that the organization has an element of sensitivity to the environment that never existed before. In doing that, the organization learns how to change, and more importantly, how to change management. This cannot be done overnight. ISO 14001 is a long-term process and through continual improvement the organization can achieve its environmental goals over time.

However, there are conceptual traps to the process of shifting paradigms. The consequence can be confusion. Organizations must be careful about what is put into the standard and what is required of employees who are volunteering to take on new responsibility. If too much is asked of volunteers, they may become disenchanted with the effort.

Does it matter when all of an organization's goals are reached? In reality, an organization should never reach that point, because it will continue to improve forever.

3

Use of Environmental Management Systems and ISO 14001 in the Public Sector

U.S. ENVIRONMENTAL PROTECTION AGENCY

Mary McKiel
Director
Standards Network

Remember that, although there is an accreditation program associated with ISO 14001, the standard does not require any kind of registration or certification. Its use is completely voluntary. If a federal agency, for example, wishes to follow ISO 14001 as the basis for management of its environmental responsibilities, then it can choose whether to go through the certification process. ISO 14001 and certification are tools for organizations to use whether the organization is a federal agency, a corporation, or an academic institution.

The number of academic institutions across the country that are using ISO 14001 not only as a facilities-based management approach, but also as an approach to developing interdisciplinary curricula, is increasing. If you look to the future and project the direction that the environmental protection paradigm will take, the question arises of how to train people to meet the challenges created by this new kind of integration. It is encouraging that academic institutions are taking on this challenge.

The U.S. Environmental Protection Agency (EPA) has been examining pollution prevention since enactment of the Pollution Prevention Act. The agency has focused on compliance and enforcement as its fundamental mission. EPA exists to enforce regulations and to provide for compliance. At some point the question had to be asked as to how the United States addresses environment-related issues that are not regulated. What about efforts to avoid causing pollution in the first place? EPA has had a strong pollution prevention program that has been growing in strength, especially for the past five years.

Into this debate comes the voluntary ISO 14001 standard. It appears to be compatible with the goals and movement toward pollution prevention, and so, EPA is evaluating its utility. In addition to ISO 14001, EPA is looking at other industry programs. It recently published an agency policy that essentially says that environmental management systems (EMSs) are good systems for organizations to use. ISO 14001 appears to have a framework for a good EMS. However, at this point, EPA will not be offering additional benefits or incentives to organizations that undertake ISO 14001 registration. EPA needs to understand whether any correlation exists between the use of ISO 14001 and the tendency of organizations to achieve or even exceed compliance or to improve their overall environmental performance. Consequently, EPA is involved in a series of pilot projects and is evaluating the results.

U.S. ENVIRONMENTAL PROTECTION AGENCY

Sarah Walsh
Project Manager
Federal Facilities Enforcement Office

The Code of Environmental Management Principles (CEMP) was developed by the U.S. Environmental Protection Agency (EPA) in response to Executive Order 12856, Federal Compliance with Right-to-Know Laws and Pollution Prevention Requirements which was signed on August 3, 1993. This order contains a requirement to establish a federal environmental challenge program.

Another component of the executive order is to have federal agencies agree to a code of environmental principles, emphasizing pollution prevention, sustainable development, and state-of-the-art environmental management programs. In addition, these agencies would have to agree to submit applications to EPA for individual federal facilities to be recognized as "model installations." Agencies are also to encourage individual employees to demonstrate outstanding leadership in pollution prevention. Upon receiving this directive, EPA formed a task force and, in October 1996, it published the CEMP.

Since then, many federal agencies have endorsed the CENT on an agencywide basis with flexibility on environmental management system (EMS) implementation at the facility level. Agencies endorsing the CENT include the Central Intelligence Agency; the U.S. Postal Service (USPS); the Departments of Commerce, Defense, Treasury, Energy, Health and Human Services, and Transportation; EPA itself, the Tennessee Valley Authority (TVA), and others. Many agencies have begun creating and developing plans that help to improve the EMSs at their facilities. Some of these plans are using the model CEMP whereas others are using ISO 14001. Still others are using a hybrid of the two.

The USPS, for example, has an integrated program in which it incorporates its EMS into its existing business plan called Customer Perfect. TVA has integrated its approach into its existing environmental management responsibilities, and EPA has incorporated many health and safety considerations into its plan. When speaking about CENT, it must be distinguished from ISO 14001. EPA does not advocate one system over the other.

CEMP

The CEMP has five principles that are also contained in the ISO 14001 standard CENT addresses compliance assurance, and pollution prevention specifically, a distinguishing difference from ISO 14001. Another difference is that CEMP is a tool and a model, but not a standard.

The first principle of the CEMP is management commitment. Agencies must make a written top-management commitment to improved environmental performance. Policies must be established that emphasize pollution prevention and the need to ensure compliance with environmental requirements.

The second principle deals with compliance assurance and pollution prevention. The agency implements proactive programs that aggressively identify and address potential compliance problem areas and utilize pollution prevention approaches to correct deficiencies and improve environmental performance.

The third principle, enabling systems, calls for the agency to develop and implement the necessary measures to enable personnel to perform their functions in a manner consistent with regulatory requirements, agency environmental policies, and agency mission.

Performance and accountability are the mainstays of the fourth principle. The agency develops measures to address employee environmental performance and ensure full accountability of environmental functions.

The fifth principle focuses on measurement and improvement. The agency develops and implements a program to assess progress toward meeting its environmental goals and uses the results to improve environmental performance.

Table 3-1 shows the correlation between EMR disciplines, the CEMP, and sections of ISO 14001.

Table 3-1 Correlation Between EMR Disciplines, CEMP, and Sections of ISO 14001

EMR Discipline	CEMP	ISO 14001 Section
Organizational structure	1	4.4
Environmental commitment	1	4.2
Formality of environmental programs	2 & 3	4.2, 4.4, & 4.5
Internal and external communications	3	4.4
Staff resources, development, and training	3 & 4	4.4
Program evaluation, reporting, and corrective action	3 & 5	4.5 & 4.6
Environmental planning and risk management	2 & 3	4.3 & 4.4

Tools for Implementing EMSs

The Federal Facilities Enforcement Office of EPA has developed three tools to assist agencies in implementing EMSs. The first is the EMS Primer, available on the internet at http://www.epa.gov/oeca/fedfac/emsprimer.pdf. The second is the implementation guide for the CEMP. A third is the environmental management review (EMR) program.

The EMS Primer, developed jointly by EPA and the Department of Energy (DOE), offers tips to an agency that is creating a plan. These tips are designed primarily to help people who are starting from the lower ranks of the organization. It has specific suggestions for those who want to "sell" the concept to an organization's management. It outlines EMS elements and explains their benefits. The EMS Primer also places EMS in the context of regulations and compliance issues, the Government Performance and Results Act, pollution prevention, and other government activities.

The implementation guide, in contrast, addresses the CEMP, describing each principle and its supporting performance objectives. It provides possible agency actions to achieve each principle, as well as a self-assessment matrix for agencies to follow in implementing the CEMP. The matrix describes stages of CEMP implementation and shows five levels of accomplishment for each performance objective.

The EMR program is an evaluation of federal facilities' environmental programs and management systems. EMRs include consultative technical assistance visits intended to identify root causes of environmental performance problems. EMRs are voluntary and often are initiated by federal agencies that request and receive their reviews. EMRs are a tool to help facility, personnel attain the CEMP and move toward the ISO 14001 standard. EMRs are not compliance-oriented assessments, audits, or inspections, nor are they pollution prevention opportunity assessments.

The EMR was originally piloted in EPA Region One, the New England area. A team comprising EPA employees and contractors would go to a federal agency and provide technical or management assistance to help the agency staff create an EMR. All 10 EPA regions are now capable of providing assistance to perform these reviews.

Many initiatives are under way at EPA for providing guidance. The Cincinnati office has put together a resource directory dealing not only with EPA and other federal agency initiatives, but with state, nonprofit, and international initiatives as well. Another resource available on the Internet gives users access to EPA's Envirosense system (http://www.epa.gov/envirosense). Anyone can access information on EMRs of federal facilities, compliance and enforcement, and technical/research and development information. (See Appendix B for a more complete list of resource documents.)

U.S. DEPARTMENT OF ENERGY

John L. Stirling
Senior Environmental Protection Specialist
Office of Environmental Policy and Assistance

There has been a gradual shift away from traditional compliance in environmental protection to a more preventive mode that adopts a systems approach and examines ways to make that system effective. Managing costs is an issue of increasing importance for environmentalists. The Department of Energy, (DOE) is now on the verge of addressing the performance of its management systems, including what it costs to implement them.

DOE has been affected by declining resources, as have other agencies, both in terms of staffing and dollars. Its operations have come under increased scrutiny from the public. DOE is expected to achieve more, improve performance, and increase competence. In addition, DOE is expected to demonstrate to the public, Congress, and state regulators that it is accountable for its publicly authorized funds and is effectively executing its mission. As a management system, ISO 14001 can help DOE demonstrate excellence in accomplishing these objectives.

DOE's basic approach to EMSs is quite simple. The department wants effective programs supported by management systems that are cost-effective and continually improving. However, DOE faces a series of challenges in achieving this. Its missions include very different types of activities. For example, DOE fosters energy efficiency; maintains strategic petroleum reserves; and produces, sells, and distributes electric power. It is also responsible for the cleanup of numerous radioactive and hazardous wastes at the nation's nuclear production sites, for on-going stewardship of U.S. nuclear materials, and for leadership in nuclear safeguards and nonproliferation efforts around the world. Supporting each of these functions, DOE also provides major scientific and educational efforts through its National Laboratories.

These missions are conducted in a highly decentralized way in dozens of unique facilities across the country. In some cases, different types of mission activities sit side by side (e.g., a laboratory and cleanup site). In other cases, related mission elements are in widely distributed locations. Further complicating the task, over 90 percent of DOE's budget is spent on competitively awarded contracts rather than salaries for federal employees. Thus DOE's approach to environmental protection must be useful to federal employees making policy and exercising oversight as well as to contractors determining how to carry out these policies. It must be an integral part of the agency's business management practices.

DOE is also responsible for relating environmental protection to protection of workers and the public. In response to a recommendation from the Defense Nuclear Facilities Safety Board, DOE has developed an Integrated

Safety Management System to address all three issues. The system is still being developed and an important challenge is to ensure that environmental protection is a strong component. An EMS, such as ISO 14001 or one based on its tenets, can be that component of the larger integrated management system.

ISO 14001 uses a systems approach, which is critical. It ensures that operations are designed to reduce pollution and continually improve. It is designed to support regulatory compliance although it is not fundamentally a compliance-based approach. It also can help to integrate environmental protection as an activity into the organization's management systems.

At DOE, environmental protection activities tend to be placed in overhead budget categories. Thus, no single entity is accountable for managing them and costs are difficult to control. An EMS approach helps organizations to define more clearly the roles and responsibilities for environmental activities, often identifying gaps and overlaps that prove costly. It also raises awareness at all levels and institutionalizes commitments to preventing pollution. If an organization integrates the EMS into its broader management systems, environmental considerations then can be linked to strategic planning and budgeting processes, and program and project management activities. This makes an EMS more rational and manageable, and links it to other established systems.

Reviews of EMSs at DOE facilities found that most already had many, if not all, elements of ISO 14001 in place. Frequently, they were lacking some elements, such as a policy statement, or maybe roles and responsibilities were not clearly delineated. Sometimes elements needed to be better integrated or strengthened. An EMS can help to improve the effectiveness and efficiency of operations while reducing costs, and helping ensure compliance because of the systems approach to identifying and reducing the impacts of its operations. This includes using performance measures, goals, and targets to reduce the impact of DOE activities, such as levels of environmental pollutants, amounts of water or energy consumed, or increases in the number of "green" products used.

DOE acquisition regulations require clauses devoted to pollution prevention and waste minimization in its contracts. The EMS is then the bridge between a management system and sustainable pollution prevention system. This is one way that pollution prevention is being institutionalized into DOE operations, using an EMS approach. The same management concepts apply to safety. Many private-sector companies are already integrating these because they are cost effective and therefore make good business sense.

DOE facilities managers have expressed a lot of interest in EMS programs. As you would expect, the approaches they have taken reflect the diversity of DOE's operations. For example, Allied Signal, DOE's contractor at the Kansas City site sought and received ISO 14001 third-party certification as part of its environmental protection strategy. At the Savannah River site in South Carolina, federal and contractor staff worked together to secure a single ISO 14001 third-party certification at a site undergoing complex restoration activities for hazardous and radioactive contamination. At the Hanford site in

Washington State, also undergoing restoration, DOE included use of an EMS in its contract documents. This has been extended further, with an EMS being incorporated into an integrated health and safety management system.

Oak Ridge National Laboratory decided to implement an ISO 14001 system within its management office. This office has five geographically noncontiguous sites, some of them in different states. Their rationale for implementing ISO 14001 is that it will set up a common management framework and improve their ability to budget for, oversee operations of, and generally better manage these five sites.

On Long Island, New York, DOE's Brookhaven National Laboratory had released tritium into the groundwater. In addition to regulatory enforcement actions, it created substantial political and public distrust. To address these issues, DOE removed the contractor, brought in a new team, and negotiated a Memorandum of Understanding (MOU) with the Environmental Protection Agency and the State of New York. Two significant elements in the MOU were the requirement for an integrated safety management system and implementation of an EMS based on the tenets of ISO 14001. The new contractor team is now implementing this system.

In sum, DOE is committed to environmental protection and to integrated safety management. However, DOE facilities managers are not required to use ISO 14001 or any other EMS. Use of ISO 14001 is voluntary. An increasing number of facility managers are evaluating — and in many cases, implementing — an EMS or ISO 14001 based on their judgment that it makes good business management sense. This is a key in the evolution of environmental protection in public agencies. DOE is expected to be more than safe, more than compliant — it is expected to be effective. ISO 14001 and environmental management systems are a powerful tool for federal managers in reaching that goal.

U.S. DEPARTMENT OF DEFENSE

Rick Drawbaugh
Representing Office of the Deputy Under
Secretary for Defense-Environmental Security

In the Department of Defense (DOD), a committee has been meeting for more than one year to determine the best approach to addressing EMSs. DOD is pursuing this issue because its analysis indicates that an EMS can pay for itself through cost avoidance. Also, on the international level, DOD operates on the premise that environmental security reinforces economic security, which in turn reinforces national security.

In determining whether to implement ISO 14001, DOD asked, "Will ISO 14001

- lower net costs?
- improve environmental stewardship?
- increase integration of environmental, safety, and health risk reductions into all operations and activities?
- enhance relations with regulators?
- change the environmental management culture from being reactive and ad hoc to a performance based focus?
- fit with existing DOD organization and mission?

After studying the issue, those in DOD's environmental community believe that the adoption of an EMS such as ISO 14001 will improve planning and program processes and facilitate informed decision making on budgeting and resource allocations.

Environmental stewardship is one of DOD's goals. An EMS provides for an integration of activities such that, if DOD plans, implements, checks, and reviews its environmental activities, it will work across the entire organization, not just the environmental, safety, and health community, but the operational community as well. Increasing the integration of environmental, safety, and health issues will mean that the people who are responsible for those activities will work together and advocate for resources together, and thus enhance environmental stewardship.

There have been positive reactions from EPA and from the states to DOD's efforts to enhance relations with regulators. Military installations that have not always been in compliance with environmental regulations will, as a result of developing and implementing an EMS, be given consideration for fewer audits.

One positive outcome of existing EMS programs is a 90 percent reduction in notices of violations in the past six years, which leads to an important point. The DOD has had an EMS in place for more than 15 years. It just has not had the discipline established within the ISO 14001 standard, and that is why facilities have not always been in compliance.

When looking at resource allocation, the culture of DOD is to focus on compliance. One DOD objective in implementing an EMS is to have a good set of requirements and to show what the fiscal impact is going to be on overall operations. Others in the organization must be convinced that resources should be allocated to pollution prevention.

In June 1997, an interim DOD policy was issued that said it would not fund ISO 14001 third-party certification without an economic analysis showing that benefits outweighed the costs. It called for the establishment of a pilot program and stressed the importance of on-going communication on ISO 14001-related activities.

ISO 14001 Pilot Study

A voluntary pilot study was established in September 1997. Each service was asked to nominate sites for the study. The installations selected will participate in a two-year pilot study (see Appendix C) that will evaluate the installation's answers to a mix of quantitative and self-assessment questions. The quantitative questions focus on funding levels, regulatory findings, number of permits, and energy consumption. The self-assessment questions focus on how the installation's current environmental program works, why the installation wants to implement ISO 14001, the benefits anticipated, and on self-developed environmental metrics and performance indicators.

The pilot study allowed installations to select their own operational or mission performance metrics. Every six months, answers to the baseline questions were updated and the costs of implementation were measured.

The last part of the pilot study was the self-developed matrix and performance standards. Each group had to develop its own objectives and targets because DOD has a variety of installation types, ranging from small installations of perhaps one person to very large installations, and test installations, product centers, depots, and air logistics centers. Each has different objectives and targets. Currently, the greatest challenge they face is deciding on their standards.

There are two issues of which one needs to be wary when dealing with pilot studies. The first is the Hawthorne effect, i.e., the fact that an organization is watching the process will result in changes being noted. The second has to do with making comparisons. There is already a management system in place. Now something new is going to be overlayed. How is the real effect determined? What caused the effect? Was it the ISO 14001-like system or was it just a matter of doing business, just a natural continuous process improvement?

Anticipated Benefits from the Pilot Sites

One anticipated benefit of an EMS will be an improved discipline with planning and programming. Another is the DOD's standardization of environmental activities. The department's intention is to have a standard approach to doing business, not to have everyone the same. Again, DOD, has very different installations with very different activities. From the international perspective, with standardization comes a set of environmental guidelines for the military sector in 44 nations, principally in Eastern, Western and Central Europe; Africa; and North America. DOD will add guidelines for EMSs in the military sector as part of a NATO pilot study.

An EMS should provide improvements in identification and correction of negative environmental impacts, as well as improved integration of environmental activities. Relations with regulators will be enhanced and so will the DOD's compliance with regulations. An EMS also should result in increased awareness and diligence by all personnel regarding the environment. DOD also hopes to see better planning and the fostering of innovation.

Further, DOD anticipates increased competitiveness among its installations. DOD's depots and logistics centers compete for work among the services and outside the military. This is an opportunity for them to improve their competitiveness and includes them in cases in which an ISO 14001 certification is required.

The true cost of implementing an EMS often is underestimated. Not only must the cost include the outside contracting fee, but also the salaries for in-house staff devoted to the project. The total costs will vary depending upon how advanced the existing EMS is. However, an EMS should result in reduced costs in the long run.

An EMS on its own should be able to convince people that it makes good business sense, even for the DOD and national defense. DOD owns the largest amount of property in the United States of any public institution. It is critical that DOD show the public that it is a good environmental steward.

In summary, DOD believes that an EMS makes good sense. It is fully supportive of an ISO 14001-like structure, but, until the results of the pilot study are seen, it is not certain to what level DOD will encourage ISO 14001 to be taken up by the services.

ISO 14001 has many potential benefits and can improve DOD's already strong environmental program. The pilot study will help to determine whether the added benefit is worth the associated cost. If indeed the benefits are worth the cost, the pilot study will help DOD make the decision whether to mandate ISO 14001, encourage it, or leave it as an option.

CALIFORNIA ENVIRONMENTAL PROTECTION AGENCY

Robert Stephens
Deputy Director
Science, Pollution Prevention &
Technology Program

Why are state agencies involved in implementing environmental management systems? One principal factor is to have an outcome-based system, one aimed at achieving better outcomes more efficiently and effectively. In many ways, public agencies at the state level are trying to redefine success to regulatory agencies, moving away from "bean counting" and a focus on compliance toward some measure of environmental performance. State environmental protection agencies have interpreted "environmental protection" over the past 20 years as counting compliance. More and more of these agencies are realizing that there are potentially much better ways to define environmental protection. They are exploring ways to measure what their missions are supposed to be in the first place, that is, environmental protection.

"Compliance" means compliance with the federal laws, a set of standards that were point-in-time concepts relating to environmental performance and protection. In some cases, they are fairly good measures and in others they are not. Those issues need to be revisited because, in fact, the mission of these agencies *is* environmental protection, environmental stewardship.

Another principal driver for state activities is credibility of the system. As EMSs, ISO 14000, and others are evaluated, along with their ability to achieve better outcomes more efficiently and effectively, credibility of the systems becomes extremely important, because state agencies are public agents. That credibility comes from open information that is valid and believable. This is not unique to public agencies, but applies to everyone who is interested in better outcomes.

Devolution of Authority to States

In the past 10 years, there has been a growing devolution of authority to the states from the federal level. New initiatives will further this trend in the environmental arena. Stakeholders, the public, and industry are demanding a better standard of performance and resource use from the states. There has to be a better way of doing business, or many state agencies will be out of business.

Foundation policy guides on which states operate include one from the National Governors' Association and another from the Environmental Council of the States. These guides basically reaffirm the notion that states are the laboratories of democracies, that they must be innovative, exploring new ways

Multi-State Working Group (MSWG)

Fifteen states are involved in the MSWG [1], a rather informal network of professionals who have a common interest in exploring the policy implications, particularly the public policy implications, related to EMSs and ISO 14000. The working group also has other stakeholders, including two federal agencies, a growing contingent of nongovernmental organizations (NGOs), and environmental representatives. A recent addition is an NGO consortium, comprising many national NGOs and the Sierra Club. An increasing number of academic institutions and companies in the private sector also are involved. These include a coalition for the implementation of ISO 14000 made up of chemical manufacturers, the American Automobile Association, American Forest Products, and a group of 15 other associations.

The initiative is clearly a federal/state partnership, with states implementing most of the federal statutes in this country. The states view themselves as innovators, and that is why they believe that it is appropriate to look at new paradigms.

Mission and Objectives of the MSWG

The mission and objectives of the MSWG are fairly focused. It is interested in developing information about the implementation of ISO 14001, EMSs, and other related systems in order to make responsible decisions about public policy. The first step, though, is to identify the key public policy, questions that need to be addressed. The next is to identify the information necessary to address these particular policy questions and then create a national database where this information could be deposited and accessed.

The MSWG will try to develop a system or a data quality structure such that the information that flows to the national database is credible and useful for answering the key policy questions. This is a principal focus of the MSWG. The source of the data is a series of pilot projects in the 12 to 15 states that are members of MSWG. These pilot projects are being conducted under a variety of models, depending upon the state; the main differences involve the roles that the public regulatory agencies play in partnerships with organizations implementing ISO 14000. In some cases, public agencies play a very passive role, approaching organizations for certain kinds of information about ISO 14000 implementation in key areas that could be placed in the national database for evaluation with other projects around the country. In other cases,

---

[1] The MSWG includes representatives from Arizona, California, Illinois, Massachusetts, Minnesota, North Carolina, Oregon, Pennsylvania, Texas, and Wisconsin, among other states.

state and local agencies are involved in not only the design of EMSs at facilities but work with companies and other organizations to define targets and objectives, perform inventories, and make significance determinations. In these cases, the system is being built in a partnership and is generating information about the performance of EMSs.

MSWG spent its first year measuring the results of implementing ISO 14001 EMSs and the types of information generated in order to identify, the key categories of information to be tracked in support of long-term public policy questions. The result was a document called the "Environmental Management Systems Voluntary Project Evaluation Guidance," published in February 1998 by the National Institute of Standards and Technology in Gaithersburg, Maryland. It contains guidance about the generation of results information in six categories: environmental performance indicators, environmental condition indicators, compliance indicators, costs and benefits, pollution prevention indicators and measures, and stakeholder preferences and confidence.

Core Information for a National Database

These are the six areas in which MSWG is interested in generating core information for the national database. The database has been established at the University of North Carolina and is receiving baseline data from some of the North Carolina pilot projects. A companion protocol document will translate guidance in the measurement areas into suggested protocols that show how to generate the information most useful to the national database. MSWG wants to collect and evaluate sufficient information, test performance-based multimedia strategies, and demonstrate streamlining options for multimedia permitting.

A critical element of these pilot projects is the interest in projects that are truly multimedia rather than single media. It is hoped that EMSs will attract not only organizations that are managing their facilities, but regulatory agencies that are designed around single media. Compliance assurance is essential to all of the pilot projects and is a given for participation in the group projects. It is important to show that high levels of compliance are achieved through implementation of EMSs.

MSWG is concerned about cost analysis. It also is interested in public disclosure and stakeholder participation, which can be controversial. To date, this has not been a problem in most projects. In fact, many of the organizations dealing with MSWG have good ideas about how stakeholder participation is supposed to function.

One of the issues that will be addressed as the group explores the interface of existing regulatory regimes to EMS is that of audit functions. Audits are designed to generate a certain kind of information. One topic to be explored in a number of states is whether there are different and more effective ways to generate that information, using some mix of internal audits, third-part audits, and government inspections.

Another issue is electronic reporting of information. When discussions are held about the ways in which regulatory agencies have generated and transmitted information for purposes of decision making, it is clear that those agencies are in the previous century, technologically. Huge amounts of money are spent on generating large volumes of data that often are of little use, winding up in archive boxes, on tables, or in file storage rooms somewhere. An organized management system can lead to much more efficient ways to generate better quality information that can be transmitted electronically.

Another issue being investigated is the area of permitting activities. Can an integrated system created around EMSs be used to change the way in which permits are issued, by whom they are issued, their lifetimes, and how they are monitored?

MSWG anticipates that there are over 100 pilot projects that are going to be conducted over the next one to two years to generate data on performance and results that will go to the national database.

Why would organizations want to participate and partner with a regulatory agency? Recognition. California, Pennsylvania, North Carolina, and other states want to recognize environmental leadership companies and organizations within the state that are willing to step forward and say there is a better way to achieve environmental protection. MSWG is talking about potentially major changes in public policy in the way the regulatory systems work. Those who partner with regulatory agencies have the opportunity to help frame the debate. They can sit at the table and influence how activities will be carried out in the future, such as how audits are conducted or how to permits are issued.

U.S. POSTAL SERVICE

John Bridges
Area Environmental Compliance Coordinator
Capital/Metro Operations

For the U.S. Postal Service (USPS), the key to success in moving beyond simply complying with environmental regulations has been to develop and implement an environmental management system (EMS) by using the standard methods approach. The system is built on three stakeholders and their interests: the voice of the employee, the voice of business and the voice of the customer. This provides an opportunity to gauge the impact of how the USPS does business and for letting the customer know exactly what the USPS is trying to accomplish.

After many years of evaluation, the USPS concluded that there are more than 160 environmental aspects of postal operations. These were divided into 11 target areas, including leadership and compliance.

The issues affecting postal operations that target management and compliance requirements are consistently changing and often unpredictable. Opportunities for improvement through EMS components are fostering continual improvement of environmental performance in management.

USPS Structure

The USPS delivers about 180 billion pieces of mail each year and employs 760,000 to 800,000 people. It manages about 40,000 facilities, ranging from small trailers in rural areas to one-million-square-foot facilities in urban areas. With 208,000 vehicles, the USPS has one of the largest fleets in the country, which also includes the largest fleet of alternative fuel vehicles.

The USPS workforce has an impact on how the EMS is implemented. An educational outreach program is being used for the line personnel to show them how a collaboration between the EMS and occupational safety and health practices could be used to improve the organization. Each of the USPS's 11 areas and 85 districts now has an area environmental compliance coordinator.

Because the environmental regulatory structure is so broad in scope, it is confusing at times. The USPS firmly believes that its environmental activities should not be driven by compliance with the law and regulations, but rather by doing the right thing for the environment. The USPS is using an EMS approach to improve environmental protection, to enhance employee commitment, and to serve its customers.

The Postmaster General signed the environmental policy, statement, which is a requirement in the EMS structure; this has given the USPS flexibility in how it handles environmental management. The opportunities for improvement are based on its Customer Perfect program, which has direct correlation to the EMS structure.

The current USPS environmental programs are viewed as a separate entity and are not integrated into USPS business decision-making processes. In addition, the system is compliance driven. The new programs are built around operations. Everything the service does has an operational impact, and operations has been subdivided into four phases. The operations element establishes a program, deploys it, implements it, and, finally, subjects it to review. This is a continuous improvement process.

The EMS Approach

The EMS is a standardized approach to environmental management that enhances operational efficiency and effectiveness. It integrates environmental issues into the USPS business decision-making process. The goal is to go beyond compliance to achieve sustainability. The USPS uses the same definition for sustainability that the President's Council on Sustainable Development has — to meet the needs of the present without compromising the ability of future generations to meet their own needs.

The EMS helps to improve the ways that the organization works, provides and improves program management, and enhances environmental performance. The EMS is integrated into business strategies and also impacts on regulatory interests and future liabilities. The goal is to be practicing environmentally sound business management in a systematic manner.

How will the USPS achieve that goal? First, through commitment. The EMS contains the policy letters and directions that define the voices of the customer, the employee, and the business. It also has national environmental performance indicators and leadership and compliance targets. All of the systems are process and results oriented, so that USPS business processes can be improved through changes in the operational delivery process by considering the environmental aspects of these processes.

The USPS has formulated a strategic plan that outlines where the organization is and how it is going to achieve success. Some processes for environmental health and safety have already been carefully evaluated. There are district and state programs and management plans. The USPS knows what is being done at each facility in terms of roles and responsibilities, operation and maintenance plans, support tools to facilitate compliance and leadership, and progress tracking.

Education and training in compliance and skills are keys to this program. The measurement and improvement system permits the USPS to reevaluate how activities are being implemented. Quality assurance reviews and baseline audits are conducted by contractors. The environmental office conducts an internal review every third year. This EMS strategy team is based on maximizing USPS strengths and eliminating weaknesses from environmental management.

Recognizing that economic, environmental, and social goals play an important role is one way to plan for the future. The USPS management believes that it provides a service to the communities it serves, and is very proactive in environmental stewardship. Some of its vehicles, for example, represent maximized efforts to address environmental concerns. The USPS is recycling waste oil and antifreeze and is recapping many used tires. Each plays an important part in how business is being done.

Similarly, the USPS is very concerned about its properties. Before a property is purchased, the service conducts a complete site assessment It has a very strong commitment to ensuring that the properties be in comparable or better condition when they are sold. Again, the goal is to go beyond compliance to achieve sustainability.

Benefits of Integrating Environmental Management

Implementation of the EMS program is showing positive results. The customers' interests are addressed through enhanced corporate image, which translates to improved market share when someone purchases stamps or other USPS products. The employees' interests are satisfied in how the USPS looks out for their welfare, especially regarding safety, health, and quality of life, that is, morale and work environment. The USPS doesn't simply tell employees what they have to do; it provides tools, whether they be personal protective equipment or information, to do the job safely.

The business interests of USPS are addressed economic gains achieved through the incorporation of programs that will enhance operational efficiency, reduce labor hours, and reduce the potential for liability.

The USPS has such an aggressive program — strategic plans and operations and management plans — that it will exceed requirements for meeting the 1998 deadlines for removing underground storage tanks, of which the USPS had thousands. It is improving customer satisfaction by developing environmental plans around operational issues. It is strengthening Postal Service effectiveness by increasing environmental awareness and education programs for employees. It is improving financial performance for environmental programs through best management practices and cost avoidance, which will better serve USPS business interests.

The bottom-line benefit of the EMS program is the economic value added to the business section of the service. Employees have a clear and powerful process to improve the bottom-line economic performance of the USPS through the implementation of an environmental management system. If the USPS looks at and deploys its plans properly, there should be a good return on investment, not only in financial gains, but also in terms of community stewardship and sustainability.

U.S. AIR FORCE

Rick Drawbaugh Deputy for Environment, Safety, and Occupational Health Technology Office of the Deputy Assistant Secretary of the Air Force for Environment, Safety & Occupational Health

When it comes to environmental management systems (EMSs), the Air Force is looking at environment, safety, and health (ESH) together. It is implementing the first policy directive within the Department of Defense (DOD) that will incorporate ESH into a single policy.

The Air Force intends to use ESH systems to improve productivity by incorporating them into core business practices. The Air Force has a business and that is national defense. Within the Air Force, business is defined as maintaining readiness, being a good neighbor, and leveraging resources. An ESH system is necessary for process-driven improvement.

What are the obstacles to establishing a quality-based environmental management system? The Air Force problem is very simple. Since 1985, budgets have been cut in half. Military staff has been cut from 602,000 to 372,000. However, the mission has not changed. The Environment, Safety, and Occupational Health Technology Office needs to better leverage resources because budgets and manpower levels will continue to go down.

Another issue is outsourcing and privatization. What impact will they have?

At present the Air Force has separate management systems or "stovepipes." There is a logistics and material support management system with its own data center, an EMS with a data center under development, and an operational readiness management system. There are separate inspections for environment, health, safety, and fire. This creates duplication of effort when the same facilities are being inspected many times. The results of having these separate systems are:

- Critical activities are not performed.
- Redundant and unnecessary activities are performed.
- Decisions are made early in the value chain without considering implications later in a process.
- People are not focused on internal and external customers.
- There is an inability to accurately cost services or products.

A single management system, at least within ESH, would reduce the overhead costs of conducting inspections, and would allow people in the field to do their jobs, which are to fly, fight, and win.

One task in putting together an EMS system is to go to the people in the field and ask them to describe their jobs. What they say and what their position descriptions say are two different things. It is difficult then to sit down and determine how to improve the situation and how to incorporate that information into a management system based on goals, objectives, targets, and, ultimately, performance measures. Performance measures are needed that relate to what those staff people actually do, not what the Air Force thinks they do.

A management system is a methodology, whether it be activity-based management or an ESH management system. To have a sound management system, tools such as performance measures, activity-based costing, and a decision support system, are needed.

Aiming to Increase Productivity

The Air Force is aiming for a 30 percent increase in productivity by the year 2005. The problem is not so much the amount of money being allocated, but how the money is being spent. Productivity is cost and performance. If existing performance can be maintained at a reduced cost, productivity can be improved just on cost alone.

To improve performance and reduce costs, customer requirements have to be defined. In the past, requirements were not identified, solutions were The individual office asked for a specific budget to solve a specific problem. This is very different from asking people what they need to do their jobs. The Air Force has made a significant effort to put in a requirements process. From there, resources must be leveraged to find the best, most cost-effective answer, and if possible, to acquire in batch rather than through single acquisitions.

The Air Force has a well-educated, experienced ESH community, with skills that can help across the Air Force. Its facilitators, enablers, processes, technology, constraints, and resources are all aimed at improving, or at least not degrading, performance at a reduced cost.

One example is a dental clinic for which an engineer identified a piece of instrumentation, a dental radiographic device that takes x rays using a digital camera. The Air Force was able to save $136,000 per year and effect a payback time of nine months. The reduced environmental costs accounted for half the savings and the other half was personnel costs. Half of a technician's time now can be used elsewhere in the laboratory. Further, for the technicians, there is a 95 percent reduction in exposures to the radiological equipment, and for patients, a 70 percent reduction. That is in only one dental clinic. The Air Force has 138 installations, so that there is the potential for significant paybacks. This type of experience needs to be duplicated.

Identifying Goals and Performance Measures

The Air Force secretariat's job is to identify goals and performance measures for ESH issues. They have to be tied together, but the key is the performance measure. If it is done properly, and if there is a sound information management system, a review, in essence, is never needed. The information available should help managers determine whether the job is being done properly or not. That is the beauty of the system. It reduces overhead costs. It reduces air staff, the military side of the Pentagon, and allows the people in the field the time to do their Jobs.

The organization's culture has to change from thinking simply about compliance. How can that be achieved if there are no performance measures to facilitate the change in the culture? Performance indicators can be separated into measures and standards. Examples of measures include effectiveness of training, no adverse press, EHS factors in financial decisions, and performance evaluations. Examples of standards are numbers of spills and releases, environmental penalties paid, number of audits, and injury frequency rate. Goals are changing but it is amazing how many managers do not know how performance measures relate back to the goal. The goal has to be there first.

One of the biggest problems that government agencies face is that they work by budget management, in which money is allocated and the agency spends it. The faster it is spent, the better chance that the agency, will get more. The difficulty with that is that it is counterintuitive to good business sense.

ESH activities, whether in business or government, generally are paid for from the overhead account. Taking these activities out of the overhead account to develop a cost center is a difficult and expensive task. How then, can one identify the costs of ESH systems?

The tools to do so exist. The Air Force is seeking to identify the old cost drivers and use them to determine the true costs to the ESH. These include the cost of materials, such as plastics, energy, electrical energy, nonproduct output and unused materials. How much of that material is purchased but never used? What is the cost of permitting and of pollutant releases? One goal is to change the way people think in terms of the amount and the kinds of materials and energy they use.

Management Approaches

The best ESH system in the world needs to be incorporated into the organization's overall management system. Then, attention must be focused on cost accounting practices. The Air Force needs to build business cases for its ESH dollars, but to do so, the costs of doing business must be known. The Air Force is moving forward with an integrated ESH policy, acknowledging that having separate policies in each of those areas is counterproductive. Last, a set of "tools" to do the job is needed.

Presently there is a "disconnect" between the polluter and the cost of pollution. If the system owners paid the bill for pollution, they would be more accountable for their actions and would improve their business practices.

Another area for improvement is to standardize accounting practices. The Air Force is looking at how industry works and is seeking a system it could adopt because it does not have the money to build a new accounting system.

The Air Force believes that it can achieve its goal to increase productivity by 30% by 2005 if it has a management system that is fully integrated into core business practices. This is not just about the ESH community, either. It includes financial, logistics, information management, and acquisition reform. All need to play a role in attaining that goal. Shifting the environmental protection paradigm from compliance to performance and cost is the ESH goal. A sound, integrated, ESH management system will provide for this paradigm shift.

4

LOCKHEED MARTIN CORPORATION'S PERSPECTIVE ON EMSS AND ISO 14001

Stephen Evanoff
Manager, External Affairs
and
Norman Varney
Associate Legal Counsel
Lockheed Martin Electronics

At Lockheed Martin, it took two years of consensus building and negotiation to get to the point where everyone recognized the value of the ISO 14001 management system that is being implemented companywide.

Lockheed Martin, like federal agencies, conducts a range of activities and operates both large and small facility complexes. It is a Fortune 20 corporation, with about $37 billion in annual sales. It has operations in all 50 states and all regions of the world, with about 15,000 of its 230,000 employees located outside the United States. It is a major contractor to the Department of Defense (DOD), the Department of Energy (DOE), and the National Aeronautics and Space Administration (NASA); it also has contracts with the United States Postal Service (USPS) and the Environmental Protection Agency (EPA). The company has an interesting mix of customers and, as a corporation, it has a unique relationship with the government, in that it is both regulated by it, and is a major service provider to it.

The Lockheed Martin corporation is best known for its work in defense, electronics, and space technologies. It also manufactures launch systems, satellites, and military aircraft. The corporation provides major information services to federal, state, and local governments and is a supplier of aircraft, space, environmental, and engineering services. In addition to operating three major DOE labs, it provides environmental remediation services internationally for DOE and also some other contract service work for EPA.

Lockheed Martin is beginning to handle rocket launches in China and Russia. It is an interesting set of relationships in which former adversaries are now business partners. The idea of transferring our knowledge about environmental, safety, and health systems (ESH) internationally to countries that do not have a strong infrastructure of such regulations is another driver for implementing the EMS companywide.

Maturity Is the Key for Large Organizations

For large corporations to understand where they need to go with ESH management systems, they must have a certain maturity and level of performance. Over the past decade, Lockheed Martin has established many systems, procedures, and pollution prevention technologies that have significantly reduced its environmental emissions. It also has corporatewide goals to further improve its safety performance.

Lockheed Martin has received a good deal of recognition from local and federal stakeholders, and some sites that have received that recognition see the value of good customer, regulator, and community relations. All that is a component of the management system that ISO 14001 embodies. Many of the company's larger and more sophisticated sites — as in the electronics sector — have many of these elements in place, which gave the company the basis to start moving toward a corporatewide structure.

The ESH vision that has been established has three elements. One is business unit self-governance. The concept is that if the business units have an integrated ESH management system in place that is rigorous and comprehensive, they will be in compliance with regulations. Compliance issues will never be out of focus; they will be managed effectively, like any other part of the unit's business.

The core business operations then will take ownership of ESH responsibilities. The need for external oversight from sector or corporate people who audit their programs will diminish. Thus, one goal is for people to see EMS as the next step in the evolution toward self-governance and cost-effective operations.

The second element of the vision is business unit performance such that the corporation has superior safety performance and continually decreasing environmental impact. That is the best way of complying — to prevent pollution in the first place.

Last, because of Lockheed Martin's size, diversity, and customer base, it must be viewed as an industry leader. The ISO 14001 system ensures that Lockheed Martin is among the companies in the forefront of this area, and ultimately that translates into a business opportunity to differentiate itself from the competition.

ISO 14001: A Rational Response

There are a number of important trends that make ISO 14001 a rational response. First, company demographics moved Lockheed Martin in the direction of having a management system that creates a common architecture across the corporation but leaves the implementation and the details to the business units. As a large corporation, Lockheed Martin has over 600 locations in the United States that have ESH significance. Over half of those facilities have fewer than 100 people on staff. Several hundred sites are so small that they do not have the dedicated staff resources to handle ESH, but they still have the requirements.

How do staff in smaller offices comply and where do they go for information? The company is creating a web-based information management system in which the corporate people can provide guidance and support information in easy-to-find and easy-to-use formats.

Second, Lockheed Martin is under continuous pressure to reduce costs and improve performance, and it became evident that putting the management system into place would help to achieve that goal. Third, the continuing enforcement emphasis at the federal level translated into a greater need for more effective compliance programs. Fourth, the federal regulatory framework was becoming stable enough so that the company was not constantly dealing with major new regulations.

Fifth, Lockheed Martin's federal government customers are becoming more knowledgeable about and sensitive to ESH issues, which are being worked into contract performance requirements. DOE contracts, for example, have numerical safety goals and performance-based contracting whereby the company can earn a bonus if its safety performance is at a specified level.

Last, the majority of Lockheed Martin's future business growth, building on its core businesses in the United States, is going to be international. Many international business opportunities that Lockheed Martin foresees in the future likely will require ISO 14001 registration.

Corporatewide Initiatives

The management system initiative is nearing completion. Lockheed Martin has revised corporate policy and procedure documents essentially to reflect the core elements of ISO 14001. An ESH management system that follows the ISO 14001 architecture is a corporatewide requirement.

The corporation has put in a management system model, a gap analysis tool (called a self-assessment or management system assessment protocol) and a risk assessment methodology. All are critical components of a web-based guidance module to assist business units in putting the management system in place. The corporate leaders feel strongly that ESH should be treated like all other business elements, where resources are allocated on the basis of risk.

Oil and gas, pharmaceuticals, and electronics firms are probably leading all industry sectors in their ESH sophistication. They are addressing ESH — quality and customer satisfaction and all other important functions — as part of their overall business planning and operational risk assessment, so that the high-risk projects take priority. Lockheed Martin is working to ensure that everyone on staff does that for ESH, with the recognition that compliance is paramount.

The company is distributing a brochure summarizing the management system to all senior managers and ESH staff around the corporation. It is a less prescriptive way to persuade people of the value of an ESH management system. The system that Lockheed Martin is putting in place is what might be called "ISO-like." It is designed to be flexible enough, like ISO 14001, to fit small, medium, and large enterprises.

There is an audit program in which the company periodically reviews compliance at the shop floor level for all facilities. A risk ranking of all sites is completed annually and the company audits are performed at facilities that are viewed as having the highest risk. Before the Lockheed Martin merger, traditional audits focused on work area compliance. Now, compliance audits are conducted, but the company is looking at things more programmatically and from a management systems standpoint. With the implementation of the corporatewide management system, the company's vision is essentially that, within the next several years, facilities will self-govern — they will audit their own compliance. External auditors, in the case of the ISO 14001 registered sites, come in every six months. Corporate management is expected to evolve to a point where it can audit the audit program (management system reviews). In other words, the company will have an external management system assessment and the goal will be to perform audits much less frequently than is currently done through compliance-based auditing. There will be less corporate involvement, fewer overhead costs, and improved emphasis on compliance, self-assessment, auditing, and self-governance at the facility level.

The company has an initiative to reduce the number of days missed due to injuries by 64 percent by the year 2000. This could save up to $40 million per year. Within the framework of a management system, it is much easier to implement these kinds of corporatewide initiatives in a large, decentralized corporation with a diverse product mix. The ISO-like model gives Lockheed Martin a common language and a common way of addressing these issues, helping people to understand the importance of this issue in the operations arena.

Lockheed Martin has a number of ISO 14001 registered facilities and two pending registrations, all within the electronics sector. Customers in Taiwan and South Korea have either directly required or inquired about Lockheed Martin's ISO 14001 and ISO 9001 registration status, which highlighted the desirability, of registration when conducting business internationally.

Lockheed Martin Electronics Sector: A Case Study

Lockheed Martin Electronics is a provider of sophisticated electronics for a variety of platforms and applications. Of approximately $8 billion in sales last year, about two-thirds was to the U.S. government.

Two factors made the electronics sector pursue ISO 14001 registration. The first factor was the very mixed constituency within the sector. The sector has 20 different corporate ancestries, such as Fairchild, Vought, General Electric, RCA, IBM, Martin Marietta, and Lockheed. They have combined into 12 domestic business units and 24 facilities with diverse approaches toward environmental aspects of the business.

The second factor was the high level of senior management sophistication about ESH issues. These individuals tend to come from large commercial enterprises such as General Electric and IBM, which have had substantial, long-term experience with Superfund, toxic tort, and media program issues. From this perspective, they see that compliance is not enough; a company can be in full compliance and still can face Superfund and toxic tort challenges. They also recognize that environmental performance is integral to business performance. There is a real appreciation of cost, productivity, and other elements of business performance of which environmental matters are an essential part. Senior management realizes, moreover, that there is a real need for individual responsibility and accountability; environmental matters of an entire business cannot be managed by a small environmental staff alone. Lockheed Martin Electronics needed to integrate environment into overall business management. It needed a comprehensive, standard management system, something that would address compliance and pollution prevention and would also facilitate continual improvement in overall environmental performance.

Achieving Registration Under ISO 14001

After some debate, Lockheed Martin Electronics decided to have all of its business units achieve registration under ISO 14001. It had the following expectations about ISO 14001:

- It would be adaptable. It would provide a baseline approach that was nonetheless open to a variety of cultures, processes, and businesses. Those business units that already had sound EMSs would be able to step into ISO 14001 relatively easily.
- It would provide a globally recognized, if not the globally dominant, environmental management standard.

- It would facilitate the exercise and demonstration of due diligence in environmental affairs, especially compliance, in a variety of regulatory and transactional contexts.
- It would distribute responsibility and ownership down to the employee level. Everyone would be responsible for the environmental aspects or elements of his or her job.
- It would allow the company to be proactive and systematic in addressing environmental matters, as opposed to being simply reactive and operating in a piecemeal fashion.
- It would help the company to reduce emissions.
- It would to reduce chemical use, waste disposal needs, and the number and scope of permits, all of which would go toward reducing costs and enhancing productivity.

The business units were allowed to choose whatever registrar they wanted, as long as it was an accredited EMS registrar.

ISO 14001 Experience

In Lockheed Martin Electronics sector's experience, ISO 14001 did prove adaptable. Regardless of corporate heritage or line of business, from making specialized chips to multiple rocket launchers to avionics repairs, ISO 14001 had the flexibility to fit every business unit.

In terms of customer preference, the company has received customer questionnaires and requests for proposals that require ISO 14001 certification, or asked for information about a business unit's EMS.

Ownership of ESH by line management and employees is growing, as responsibility for environmental management is distributed through the ranks. Business units are moving from total reliance on environmental specialists, to all personnel taking responsibility for the environmental aspects of their jobs.

ISO 14001 has proven relatively inexpensive. Larger, more sophisticated businesses were further along in developing EMSs; although these systems often were relatively informal, it took a comparatively minor effort to get ISO 14001 registration. Smaller facilities tended to be further behind in development of EMSs of any kind. For them, registration was somewhat more expensive.

Early evidence indicates that internal compliance audits show a 30 to 60 percent improvement following ISO 14001 registration. Environmental performance has improved, some business units are seeing reductions in the number and the scope of permits, and cost reductions are being observed in association with reduced regulatory applicability. It appears that productivity will be enhanced due to reduced energy, water, and material expenditures.

Although Lockheed Martin Electronics has not received any regulatory relief as a quid pro quo for ISO 14001 registration, it anticipates realizing reduced regulatory exposure through the reduction or elimination of regulated activities as a result of setting and realizing objectives and targets under ISO 14001.

Lessons Learned

Among the lessons learned from this experience is the importance of looking at the big picture. One of the benefits of ISO 14001 is that it tells an organization to survey all of its environmental aspects, to look at what is being emitted, but also at the resources being used: Is the organization using too much water, too much energy, too many materials, or too much of certain kinds of materials? It also asks that the organization pay attention not just to the environmental aspects of its manufacturing operations, but also to the environmental aspects of all services and other operations.

Another lesson learned is that ISO 14001 is a comprehensive toolbox. But, like a toolbox, it is not useful unless the organization has the appropriate tools (programs) in it and uses them.

ISO 14001 requires good faith to effectuate change. This is one of the positive aspects of having surveillance audits every six months or so: They support the exercise of such good faith.

ISO 14001 is a catalyst. It creates sensitivity to the environmental ramifications of what people do. Engineers are looking at the environmental aspects of design and being more innovative because there is greater overall sensitivity to the environmental implications of what people do. Procurement groups are asking what they can do to improve chemical management through chemical acquisition.

The overall lesson learned is to give the system time. It is too soon to say whether ISO 14001 is a success or not. It will take a couple of years to see how things develop, but from the early, often anecdotal, evidence, it appears to say positive move.

Appendix A
Biographical Sketches of the Speakers

JOHN H. BRIDGES III is an Area Environmental Compliance Coordinator for Capital Metro Operations of the U.S. Postal Service. Mr. Bridges has over 23 years of environmental and safety management experience with the Department of Defense and private industry. He is responsible for coordinating and harmonizing environmental aspects relating to postal operations within the District of Columbia, Maryland, and Northern Virginia. He is a member of the U.S. Technical Advisory Group to ISO Technical Committee 207 for ISO 14001 and participates on the President's Council on Sustainable Development.

JOSEPH CASCIO is Vice President of Environmental Management Systems of the Global Environment & Technology Foundation (GETF), a 501 (c) (3) not-for-profit organization that fosters innovation by uniting the environment, technology, and enterprise for sustainable practices throughout the world. Mr. Cascio is also the Chairman of the U.S. Technical Advisory Group for Technical Committee 207.

Mr. Cascio has been the lead U.S. delegate to the International Organization for Standardization (ISO) on the ISO 14000 environmental management standards since 1991. He previously served as the Chairman of the U.S. Technical Advisory Group for the Strategic Advisory Group on the Environment (SAGE), the precursor of ISO Technical Committee 207 (TC 207) When SAGE was superseded by TC 207, Mr. Cascio was again elected chairman of the new U.S. Technical Advisory, Group to TC 207, the position he holds today. He is recognized in the United States and throughout the world as an expert on environmental management and as the key architect and strategist in formulating the U.S. posture on ISO 14000.

Mr. Cascio coauthored *ISO 14000 Guide* published by McGraw-Hill and edited *The ISO 14000 Handbook* published by CEEM Information Services. He has authored over 24 articles and papers on environmental management, delivered over 300 speeches and presentations, and testified before congressional subcommittees on the subject.

COLONEL RICHARD B. DRAWBAUGH is Deputy for Environment, Safety, and Occupational Health Technology, Office of the Deputy Assistant Secretary of the Air Force in Washington, D.C. As deputy, he serves as the lead for developing the technology investment strategy, environment, health and safety management systems, and international affairs. The office is responsible for overseeing worldwide Air Force environment, occupational safety and health, fire prevention and protection, air base performance and operability, and interagency and intergovernmental coordination matters. He also has extensive involvement in a wide range of operational matters such as weapons system maintenance and design.

Colonel Drawbaugh was commissioned in the Air Force upon graduation from the University of Pittsburgh and has served on active duty for 20 years. He served as Chief Toxicologist in the Epidemiology Division of the School of Aerospace Medicine and later as the first toxicologist in the USAF Occupational and Environmental Health Laboratory. From 1983 until 1986, he was both the Director of Aerospace Sciences at the European Office of Aerospace Research and Development (London, UK) and Visiting Scientist to the Chemical Defence Establishment, Ministry of Defence, Porton Down, Wilshire, UK. He then returned to the United States and became Chief of Toxicology Branch, Toxic Hazards Division, at Wright-Patterson AFB, Ohio.

From 1988 to 1989, he was the director of Crew Systems Integration at the Aeronautical Systems Division, Wright-Patterson AFB, where he began to incorporate environment, safety, and occupational health considerations into weapons system design. He followed this assignment to become the Director of Environment, Safety, and Occupational Health (ESOH) at the Human Systems Center, Brooks AFB, Texas. In August 1995, Colonel Drawbaugh left San Antonio, Texas, to assume his current position in the Air Force Secretariat.

STEPHEN EVANOFF is Manager, External Affairs, with Lockheed Martin Corporate Environment, Safety, and Health (ESH) in Denver, Colorado. His primary responsibilities include developing ESH initiatives with customers, regulators, and industry associations. He is the corporation's overall coordinator for ESH issue advocacy and for technology transfer. Mr. Evanoff also serves as the Chairman of the Board of the International Cooperative for Environmental Leadership (ICEL) and represents Lockheed Martin on other international organizations and EPA committees, including the ISO 14000 US TAG, ANSI-RAB ISO 14001 EMS Council, and the EPA Subcommittee on Clean Air, Energy Conservation, and Climate Change.

His prior assignments include Manager of the Lockheed Corporate Environment, Safety and Health Program in Las Vegas, Nevada; Manager of the Environmental Resources Department at Lockheed Martin Tactical Aircraft Systems in Fort Worth, Texas; and Project Engineer in the Tactical Aircraft Systems Engineering Department, where he developed pollution prevention technologies for the F-16 Program. Mr. Evanoff served 10 years in the U.S. Air Force and received the Air Force Commendation Medal. He holds a B.S. and an M.S. in chemical engineering and is a distinguished graduate of the Illinois Institute of Technology and the Air Force ROTC Program. He is a Registered Professional Engineer, a Diplomate of the American Academy of Environmental Engineers, and a Registered Environmental Manager.

Mr. Evanoff has authored or coauthored 28 environmental papers and book chapters and holds one patent. He received the EPA Stratospheric Protection Award in 1992 and the EPA Best-of-the-Best Stratospheric Protection Award in 1997.

DR. MARY C. McKIEL is the Director, U.S. Environmental Protection Agency (EPA) Standards Network. As Director, she coordinates agency use of nongovernment standards and manages EPA's participation in the U.S. Technical Advisory Group (TAG) for the development of ISO environmental management standards. She is also the Vice Chair, U.S. Technical Advisory, Group to ISO Technical Committee 207 for Environmental Management Standards, and Vice Chair of the U.S. National Accreditation Program for ISO 14000.

In 1993, Dr. McKiel joined EPA in the Office of Prevention, Pesticides, and Toxic Substances. She was appointed to develop and manage the EPA's first cross-office program for voluntary standards.

From 1982 to 1993, Dr. McKiel served in the Federal Supply Service of the General Services Administration (GSA) as Chief of Engineering and Standards Policy, Director of Quality Standards. and First Director of the GSA Environmental Planning Program. She instituted and managed quality control and assurance programs for GSA. She also developed and implemented a federal program on acquisition of recycled and recyclable products, and developed and published GSA's first "green" catalog.

Dr. McKiel began her federal career in 1976, as an analytical chemist at the National Archives and Records Service (now an Independent Administration), where she developed chemical methods for restoring and preserving textual and nontextual materials. As a member of the U.S. Group to ISO Technical Committee (6) on Paper, she participated in developing standards for archival quality paper. Her education includes advanced degrees in Chemistry, Physics, and English.

DR. ROBERT D. STEPHENS is the Deputy Director for Science, Pollution Prevention, and Technology Program in the Department of Toxic Substances Control, California Environmental Protection Agency. In this capacity, Dr. Stephens is responsible for all scientific, engineering, and technology programs in the Department. Prior to this assignment, he was chief of the Hazardous Materials Laboratory (HML), in the Department of Toxic Substances Control. In this capacity, Dr. Stephens administered, organized, and directed the activities of HML in support of California's environmental programs. In addition, Dr. Stephens conducts research and has published over 50 papers in several areas of environmental chemistry.

For the past 15 years, Dr. Stephens has been active in the area of the relationship between environmental science and environmental policy. Dr.

Stephens is a member of the U.S. Technical Advisory Group (U.S. TAG) representing the California EPA, and he is active in several of the U.S. TAG subcommittees and working groups. Dr. Stephens chairs the California EPA task force responsible for developing the policies and programs on how the ISO 14000 standards relate to regulatory and other public policies. He has been active in the promotion of interstate cooperation in the development of ISO 14000-based public policies.

For the past three years, Dr. Stephens has been involved with the effort to develop a national environmental laboratory accreditation program and served as the first Chair of the National Environmental Laboratory Accreditation Conference.

Dr. Stephens holds a Ph.D. in chemistry from the University of California. He has been with the State of California since 1974. Prior to joining the State, he held positions in academia and industrial research.

LARRY STIRLING has been a Senior Environmental Protection Specialist with the Office of Environmental Policy and Assistance, U.S. Department of Energy (DOE), since 1988. He is a leader of the environmental management systems team and a member of TC 207, ANSI-RAB EMS Council, and Co-chair of the Federal Interagency Work Group on Environmental Management Systems. Mr. Stirling was responsible for revising the DOE environmental protection directive and worked on contract reform and directives streamlining.

Prior to his current position, Mr. Stirling was an Environmental Protection Specialist with the U.S. Environmental Protection Agency from 1980 to 1988. He also has worked as an environmental planner/policy analyst with regional and local governments in New York State and New Hampshire. Mr Stirling holds a B.S. from the State University of New York in Albany and an M.S. from Rensselaer Polytechnic Institute.

NORMAN A. VARNEY, JR., is Associate General Counsel — Environment, Safety, and Health for Lockheed Martin Corporation's Electronics Sector. Mr. Varney's legal counsel and representation includes advice on developing, coordinating, and facilitating the environmental, safety, and health initiatives of Lockheed Martin Electronics. Mr. Varney was named to his present position following the merger of Lockheed Martin Corporation and Martin Marietta Corporation in March 1995. He had been Counsel — Environment, Health, and Safety with Martin Marietta since April 1993.

Mr. Varney began his career with Citicorp counseling on international matters. He joined General Electric (GE) Company in 1979, where he progressed through a variety of increasingly responsible positions focused primarily on environmental and international law. In 1987, he joined General Electric Aerospace with responsibility for internal investigations and external representation regarding government procurement and business practice concerns.

Mr. Varney returned to environmental work in 1990, when he was appointed general counsel of GE's air pollution control systems business. In 1992, he rejoined GE Aerospace to handle environmental, safety, and health issues associated with business restructurings, including the disposition of GE Aerospace to Martin Marietta the following year.

Mr. Varney received his bachelor's degree (magna cum laude) from Duke University, and his law degree from the University of Connecticut.

SARAH E. WALSH has worked for the U.S. Environmental Protection Agency since 1991. She has served as a manager in EPA's stationary-source air compliance program and is a project manager for the Federal Facilities Enforcement Office in the Office of Enforcement and Compliance Assurance. Before joining EPA, Ms. Walsh worked as the supervisor for air quality enforcement for the State of Minnesota's Pollution Control Agency, and as a marine inspector, pollution and casualty investigator for the U.S. Coast Guard's marine environmental protection program. Ms. Walsh retired from the Coast Guard Reserves in 1995 with the rank of Commander (O–5); she had 22 years of service including 13 years Active Duty.

Ms. Walsh recently served on a one-year assignment to the Assistant Secretary of the Army's Occupational, Safety, and Environmental Health Office at the Pentagon. She was detailed to different locations to assist with special projects — preparing an Environmental Assessment for Fort Belvoir's marina development; developing a pilot study for Environmental Management Systems/ ISO 14001 at the Army's Environmental Policy Institute; advising the Army about its vessels and military watercraft to conform with the Uniform National Discharge Standards; evaluating management plans and staffing requirements for the U.S. Army Corps of Engineers' Office of Safety and Occupational Health.

Ms. Walsh has a B.S. from Towson University and Masters' degrees from Pepperdine University and the University of New Orleans in management and program evaluation.

Appendix B
Assistance and Resource Documents

Audit Policy Interpretive Guidance

Catalogue of Federal Agency Environmental Compliance/ Management Documents (EPA 300-B-94-012).

Design Guidelines for Environmental Auditing at Federal Facilities (EPA 300-B-96-011).

EMS Primer for Federal Agencies (Spring 1998)

Environmental Management System Benchmark Report: A Review of Federal Agencies and Selected Private Corporations. (EPA 300-R-94-009).

EPA Audit Policy Update (EPA 300-N-97-00 1).

Executive Guide to Facility Environmental Management by the Civilian Federal Agency Task Force

Federal Facilities Sector Notebook: A Profile of Federal Facilities (EPA 300-B-96-003).

Generic Protocol for Conducting Environmental Audits of Federal Facilities (EPA 300-B-96-012).

Implementation Guide for the Code of Environmental Management Principles for Federal Agencies (CEMP) (EPA 315-B-97-001).

Resource Directory for Environmentally Preferable Landscaping for Federal Facility Managers (EPA 300-B-96-014A, B, C, D).

Strategy for Improving Environmental Management Programs at Civilian Federal Agencies (EPA 300-B-96-006).

Appendix C
DOD Pilot Study EMS Installations

U.S. ARMY

- United States Military Academy, New York
- Tobyhanna Army Depot, Pennsylvania
- Letterkenny Army Depot, Pennsylvania
- Fort Bliss, Texas
- Radford Army Ammunition Plant, Virginia
- Fort Lewis, Washington

U.S. NAVY

- Whidbey Island Naval Air Station, Washington

U.S. MARINE CORPS

- Camp LeJeune, North Carolina

U.S. AIR FORCE

- Robins Air Force Base, Georgia
- Sheppard Air Force Base, Texas
- Eglin Air Force Base, Florida